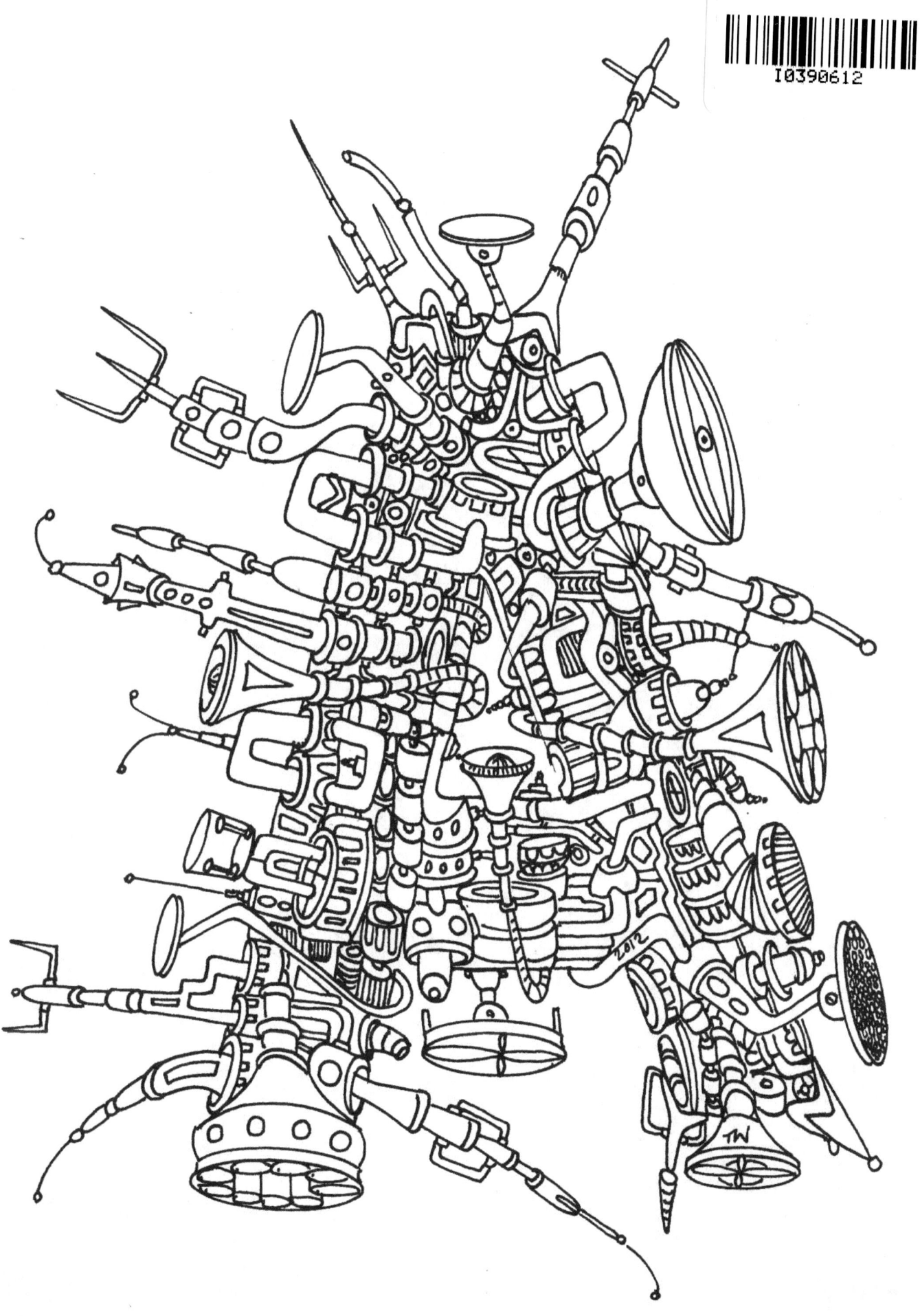

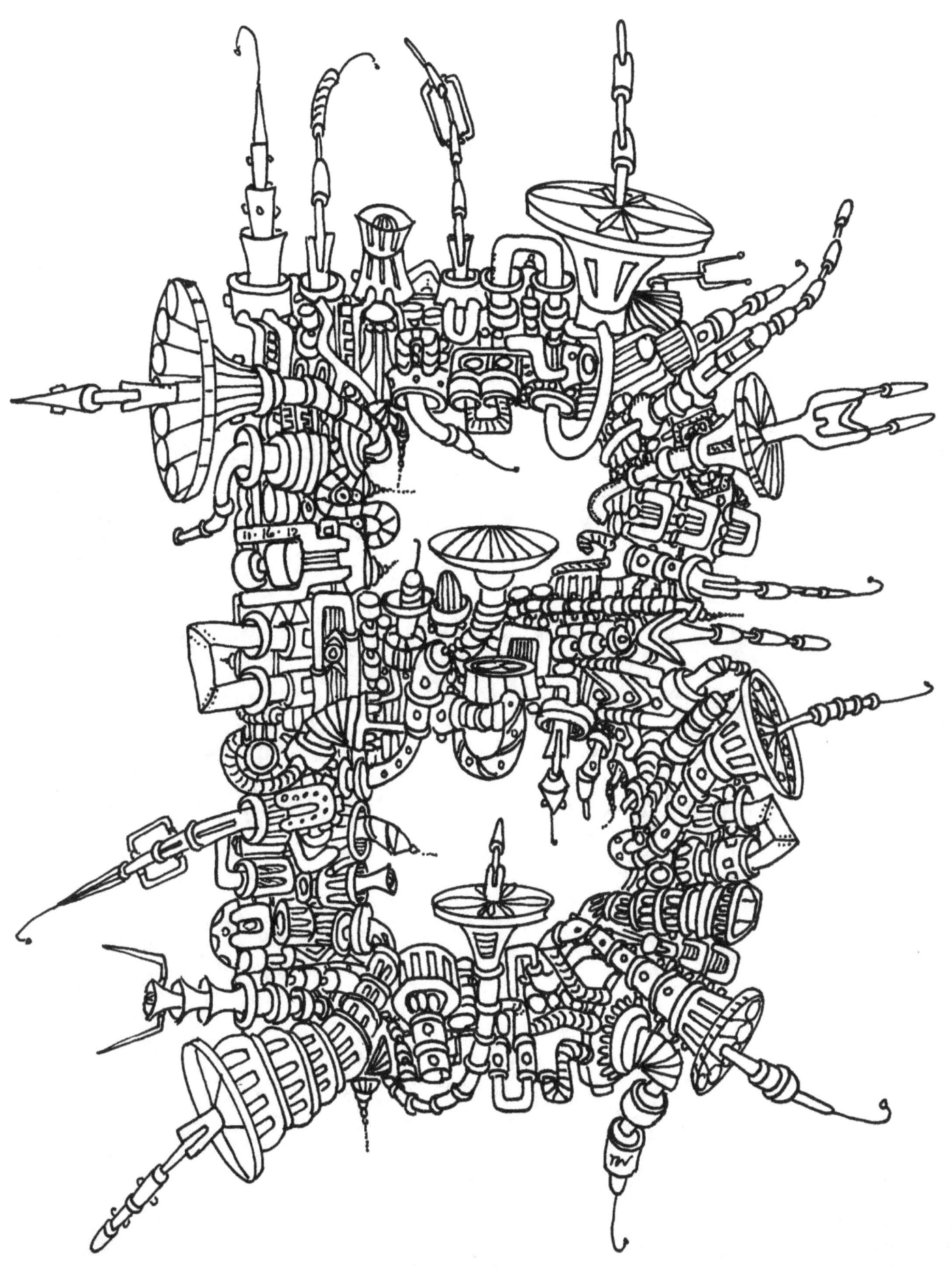

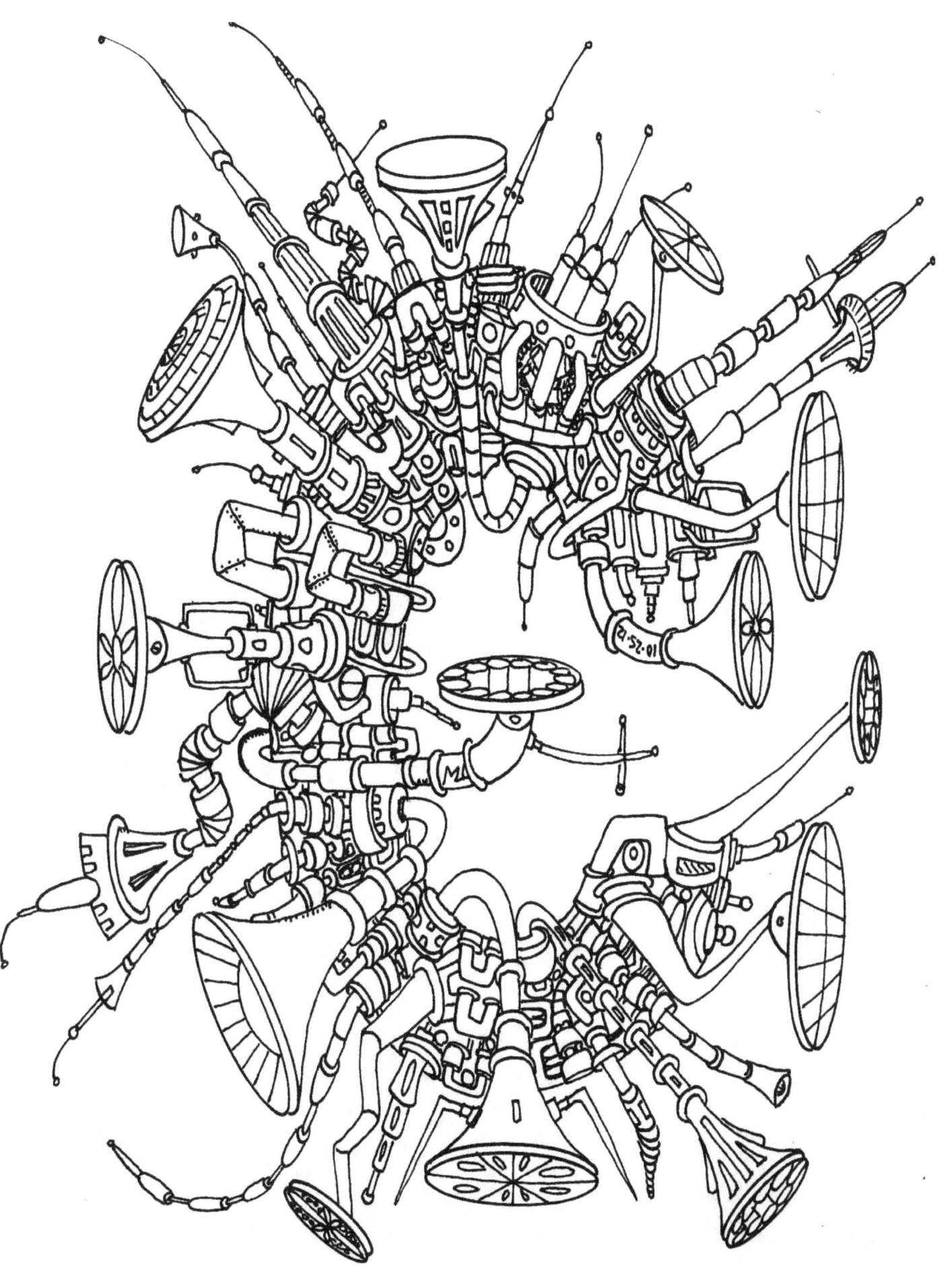

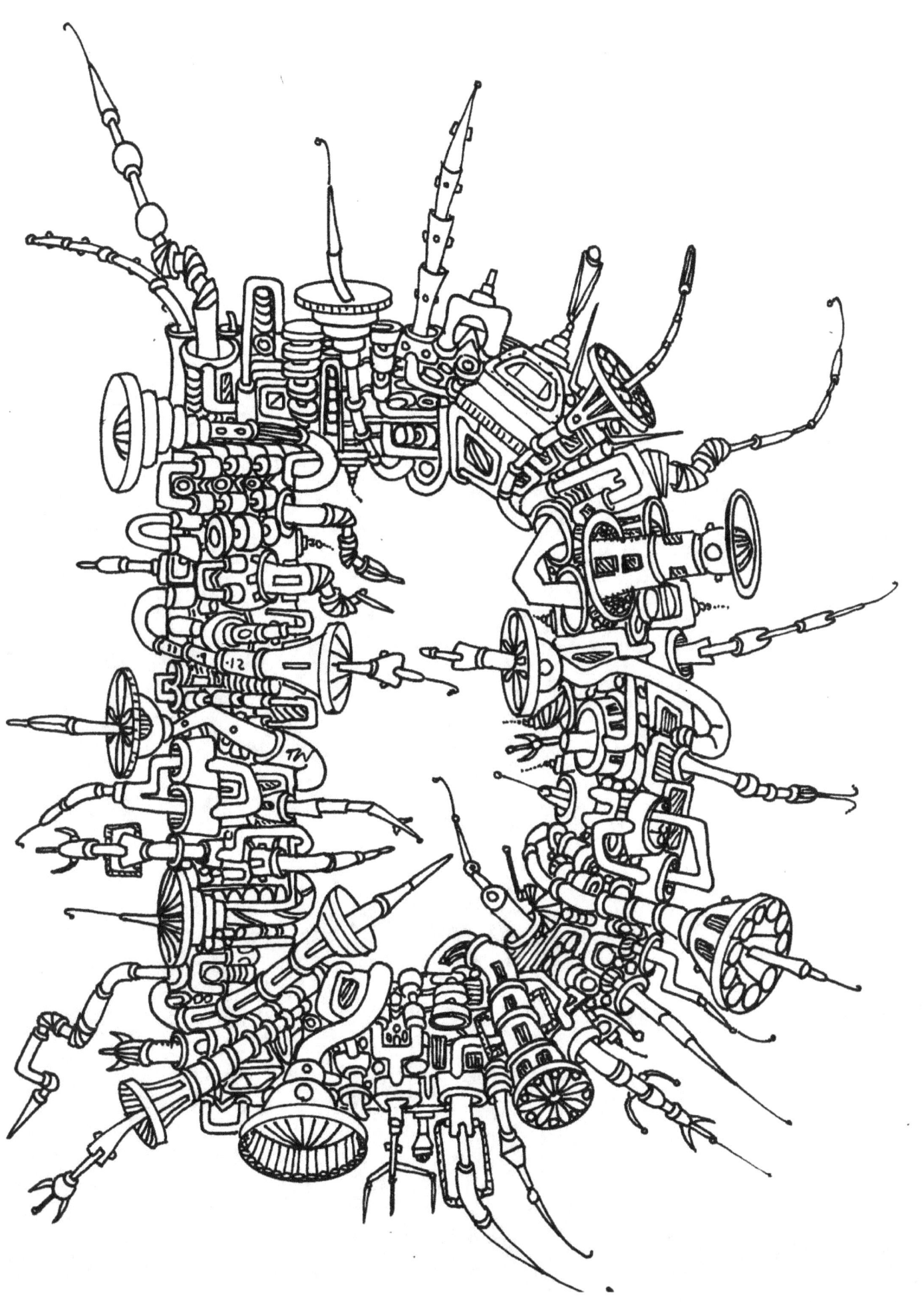

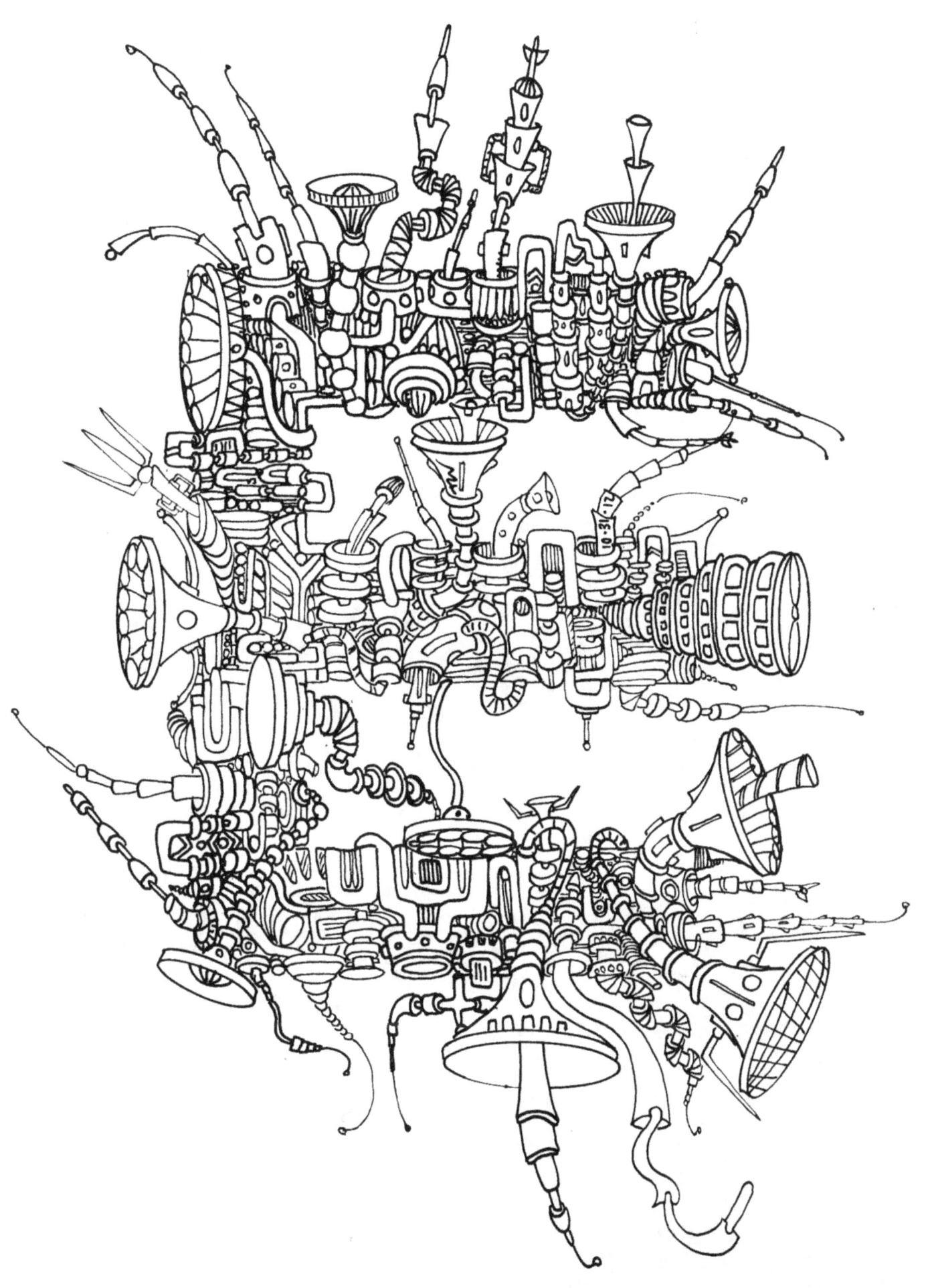

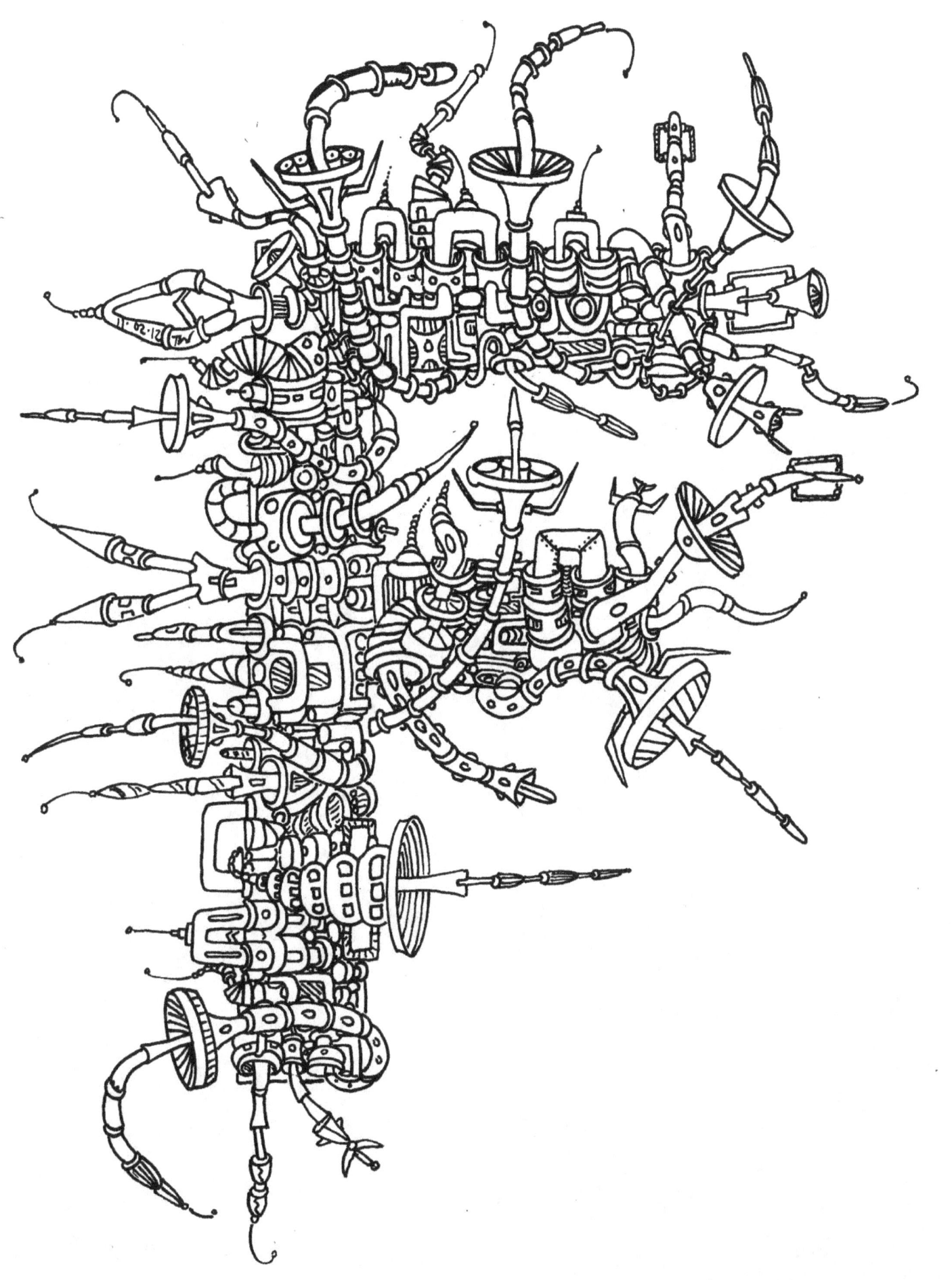

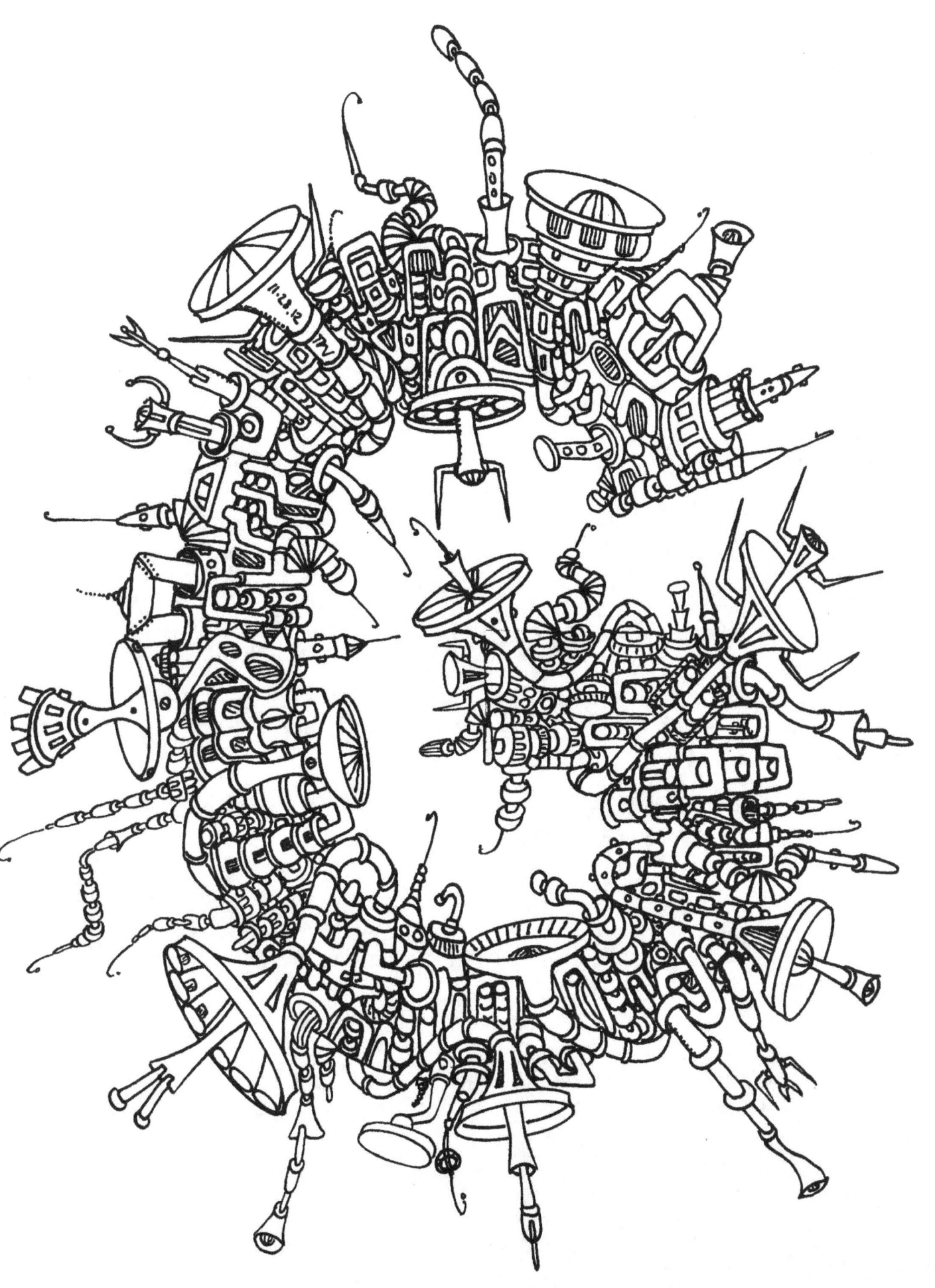

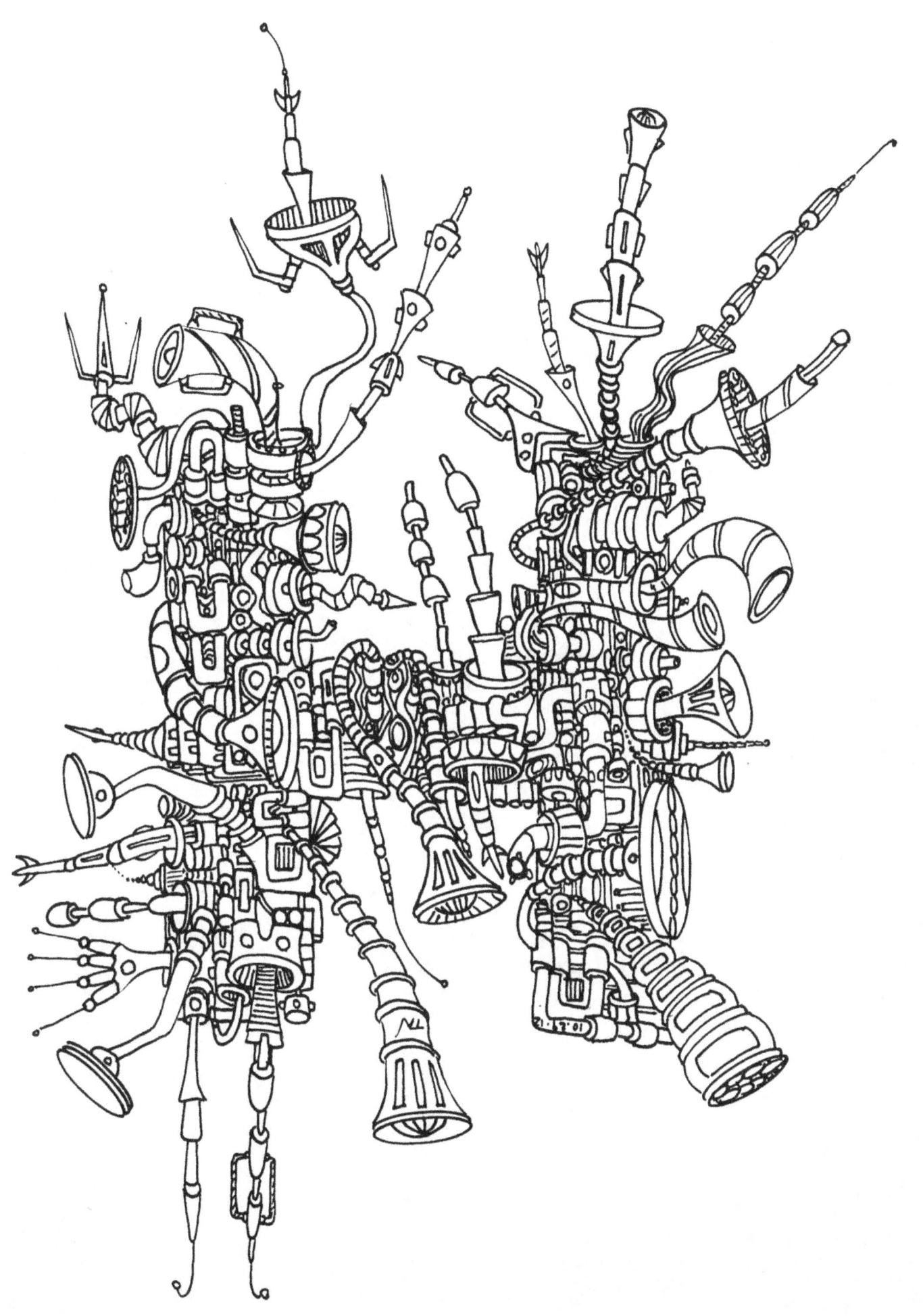

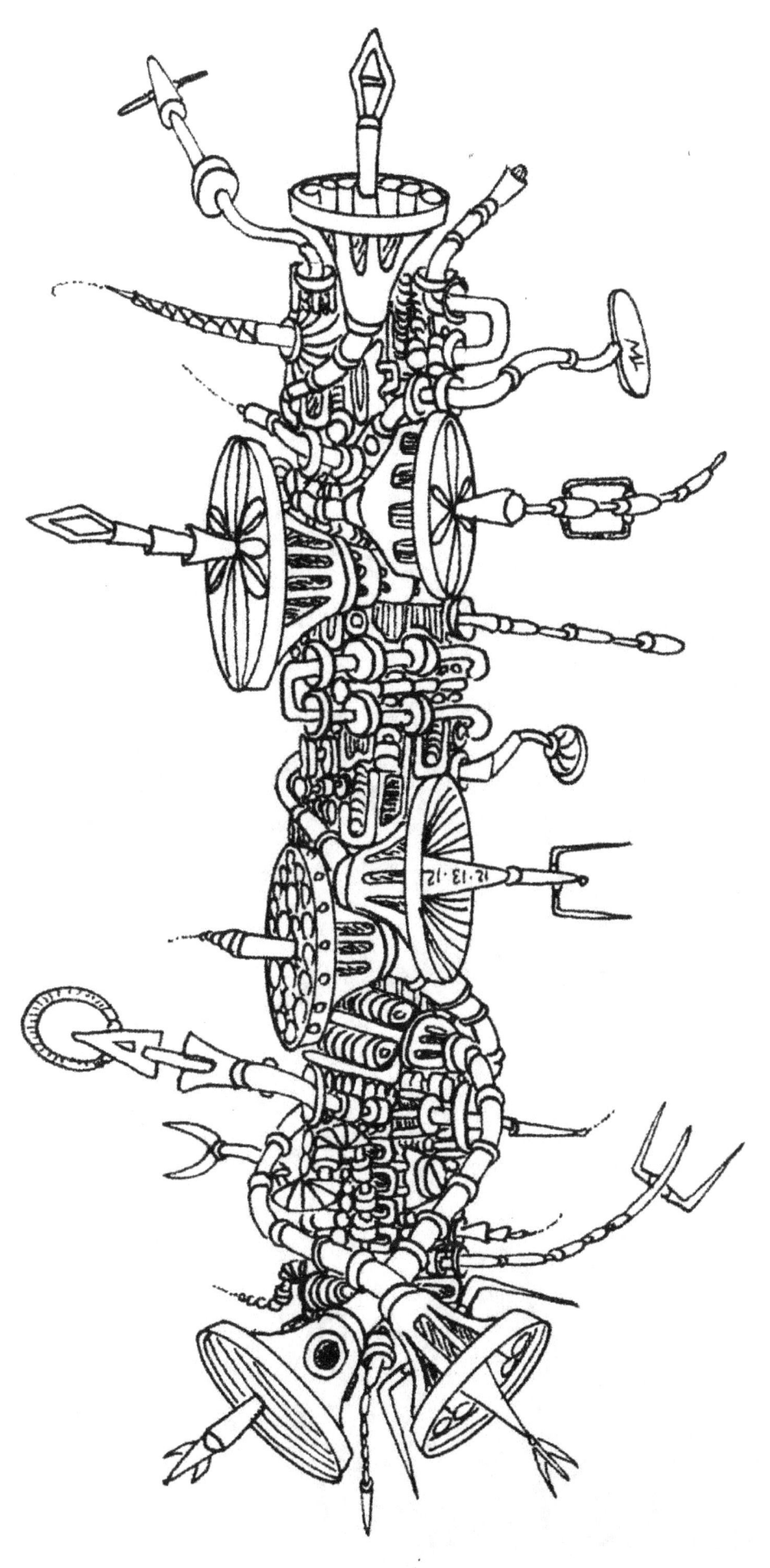

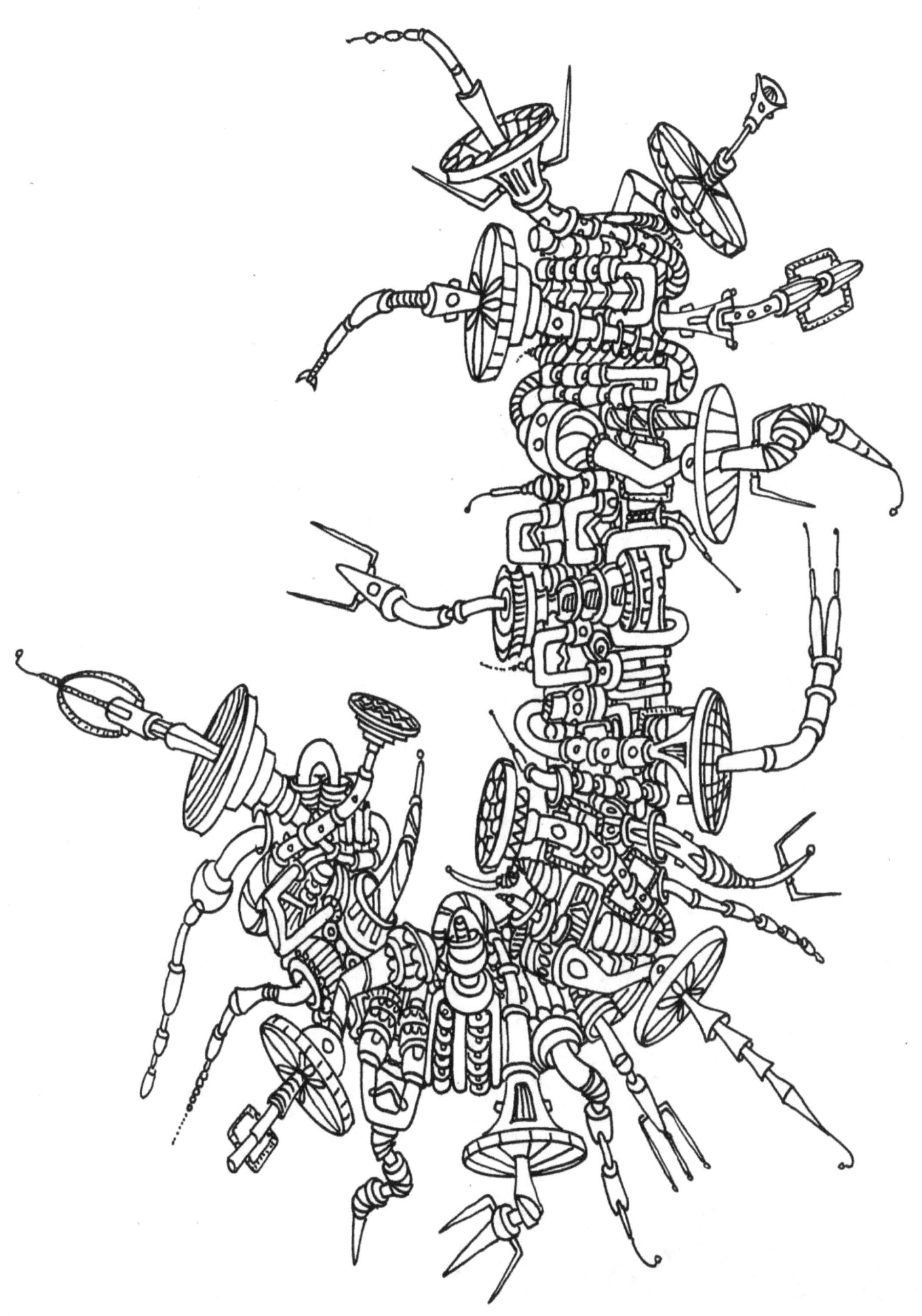

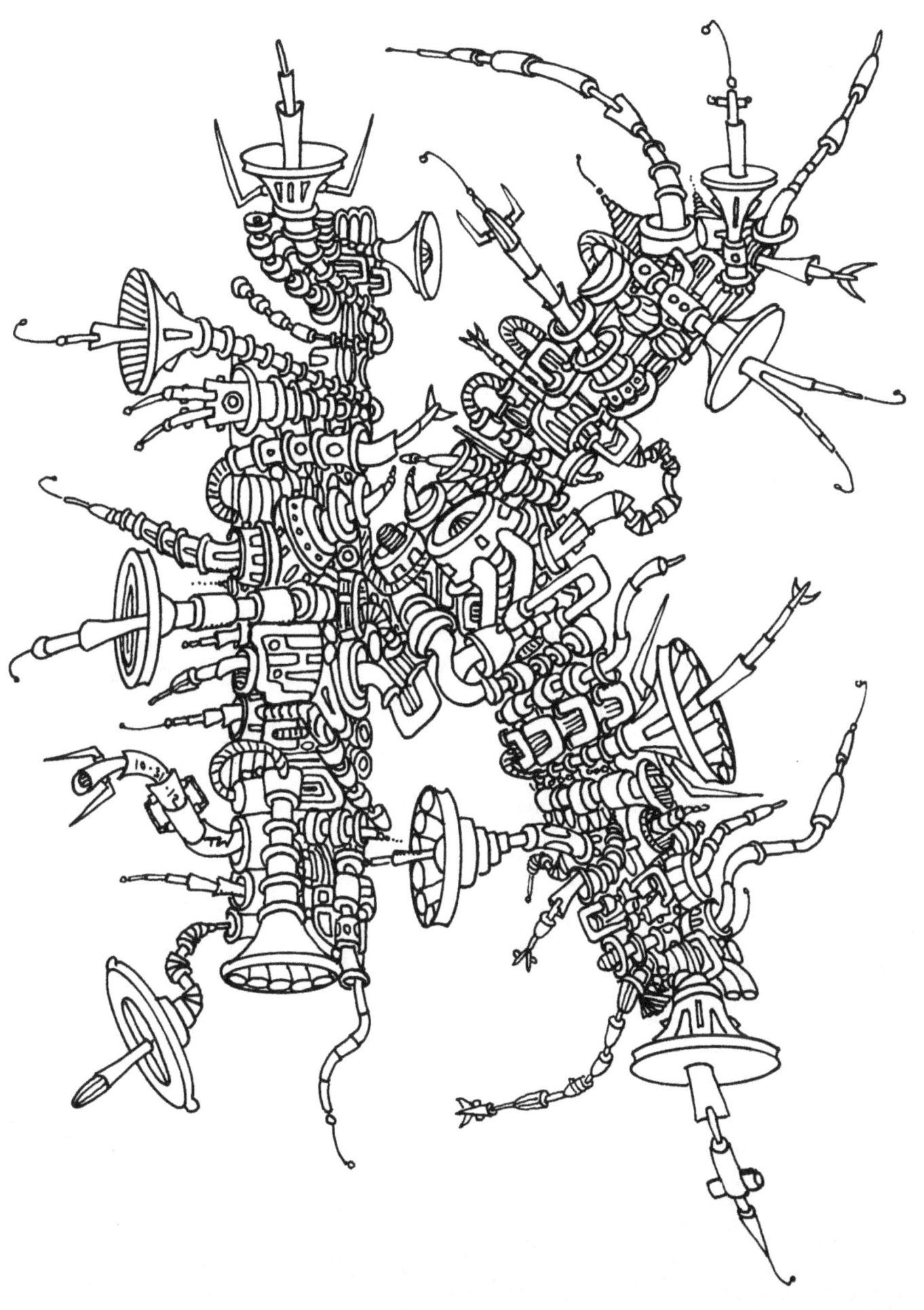

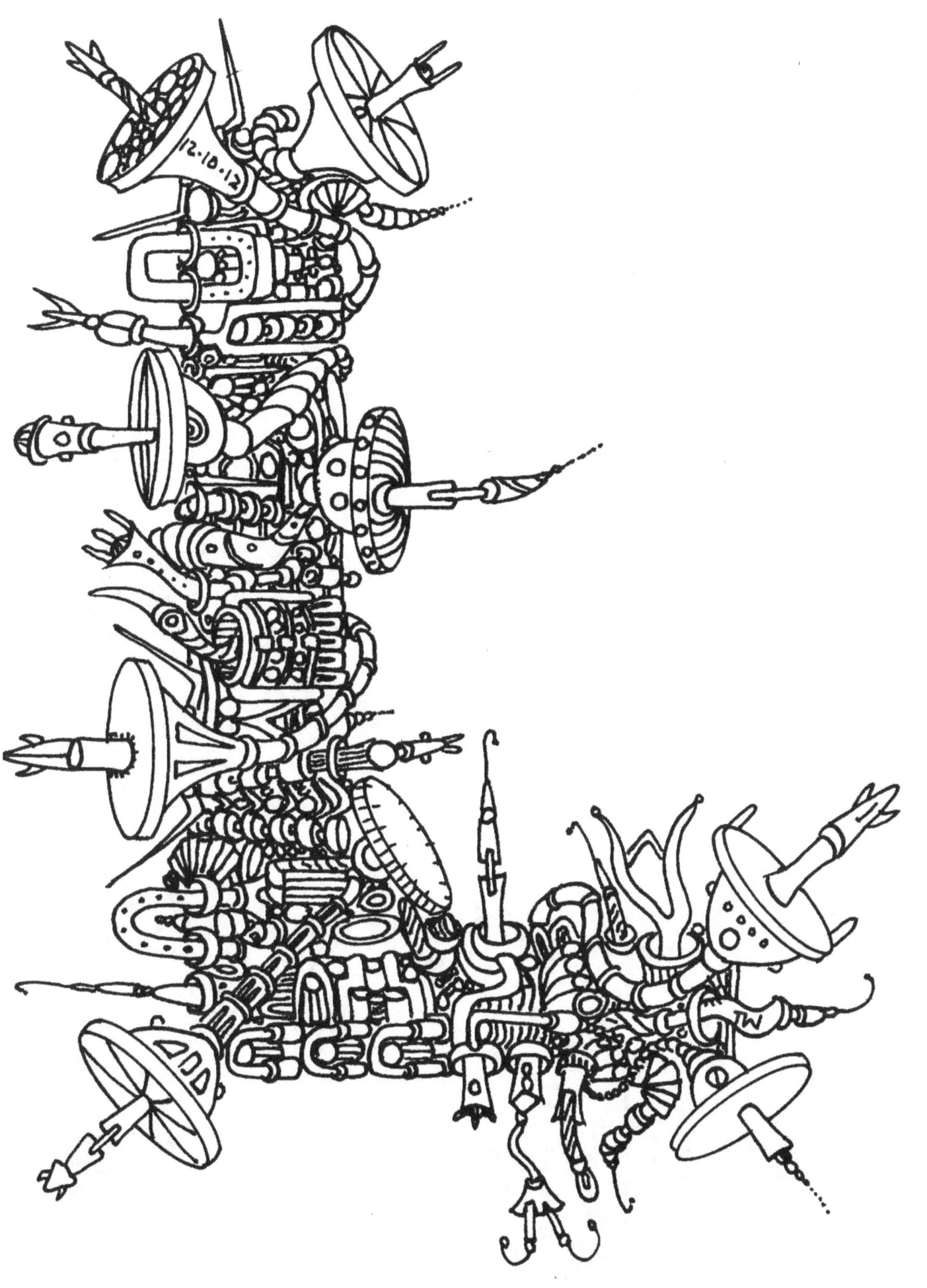

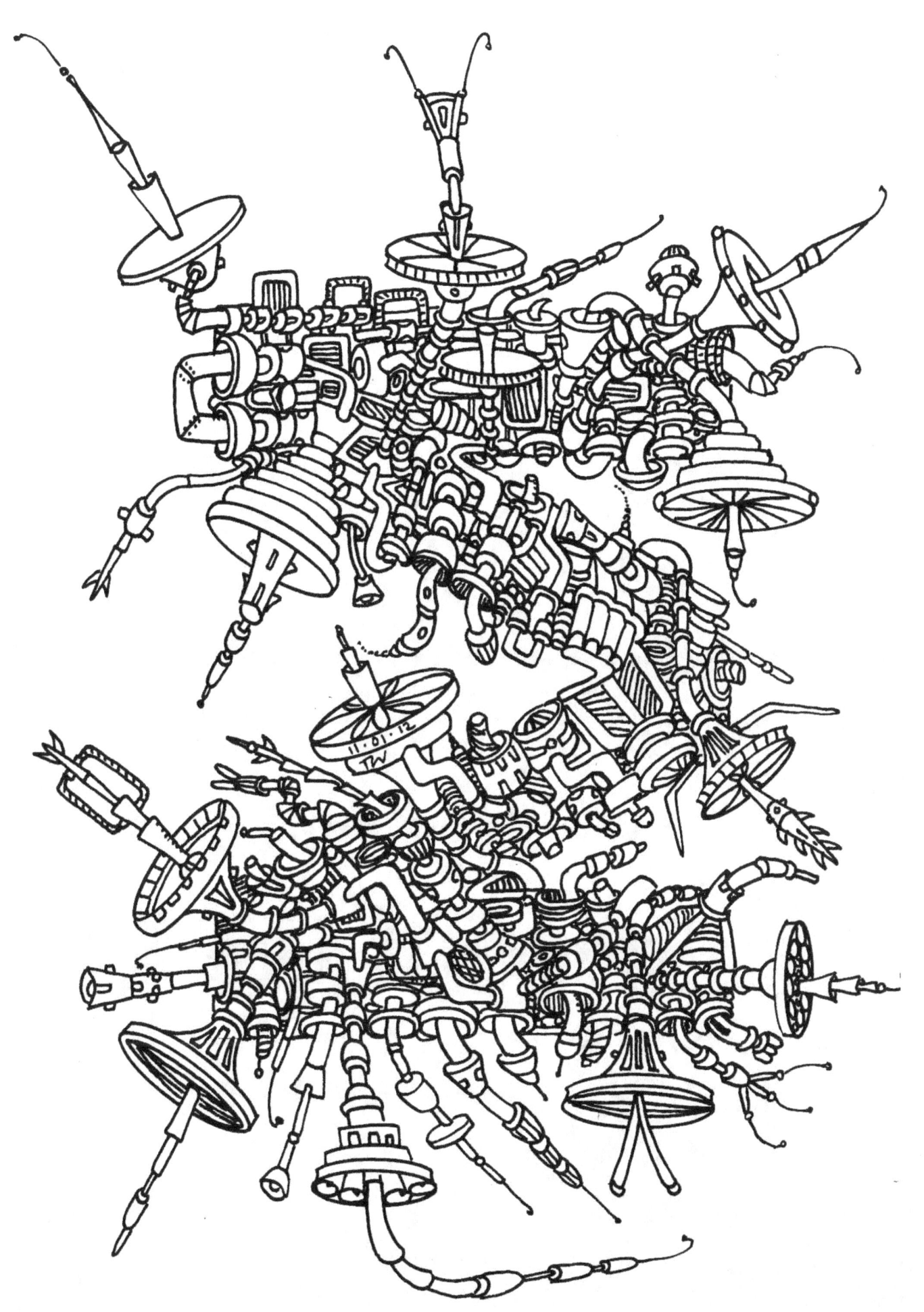

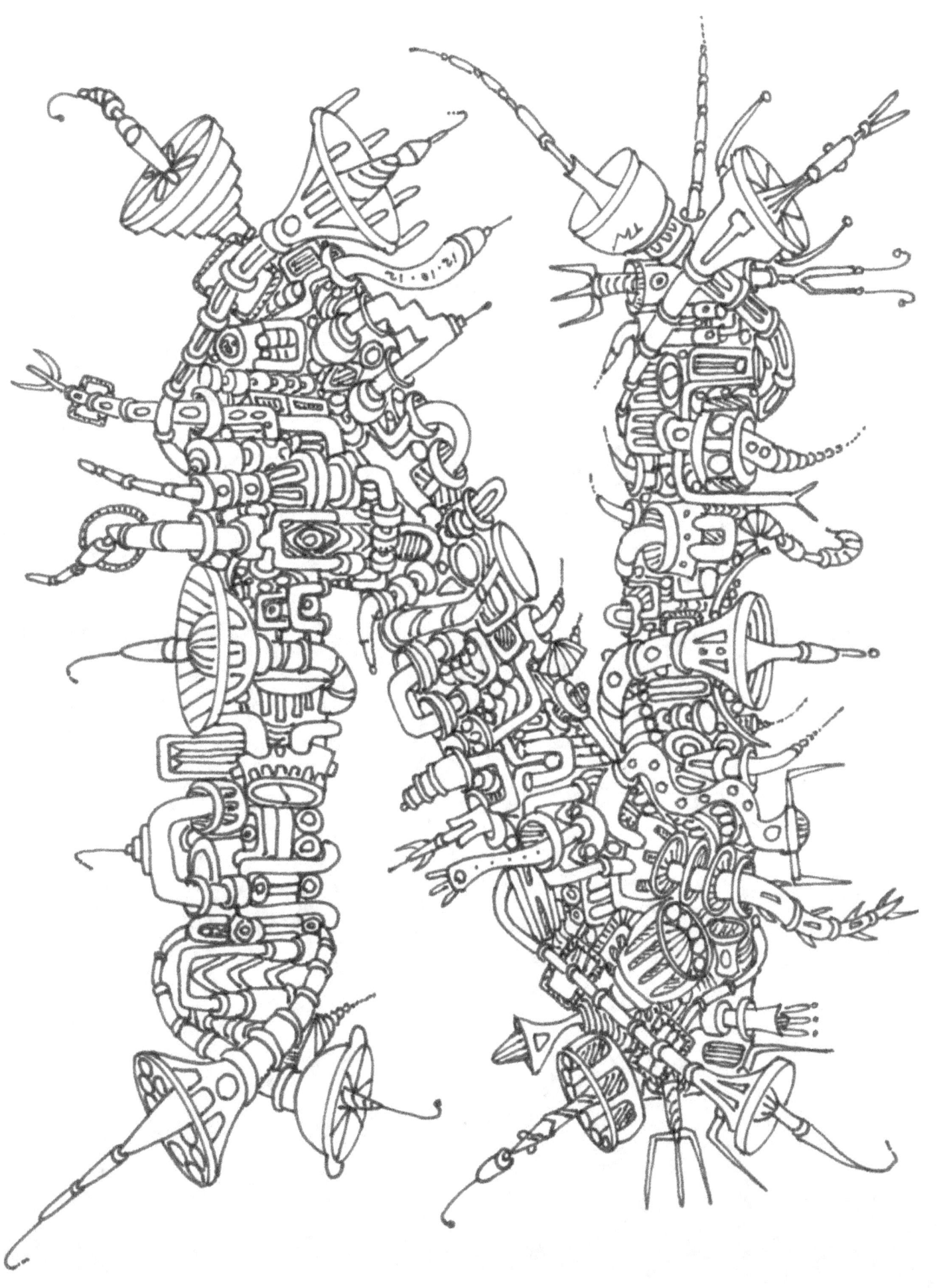

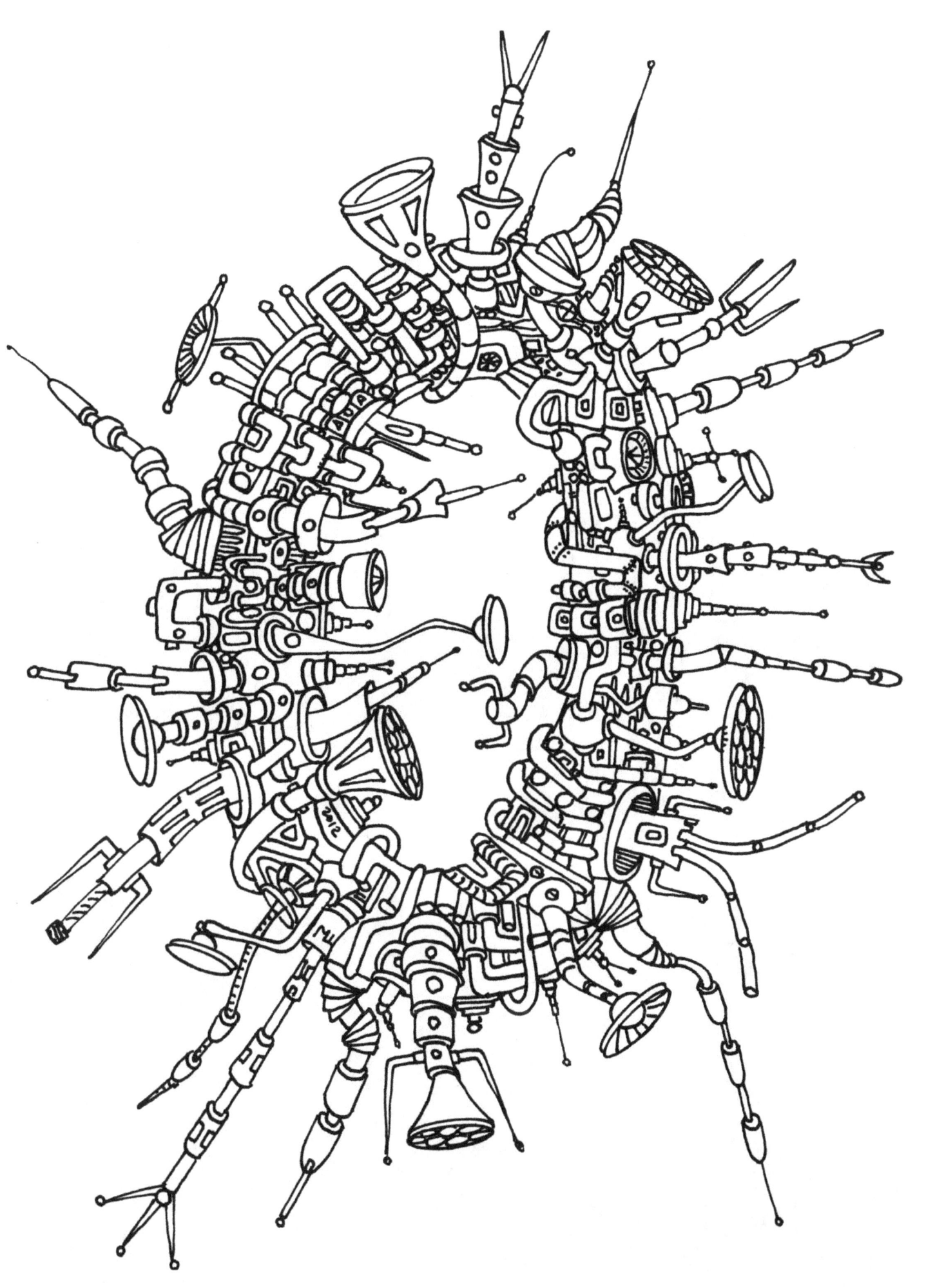

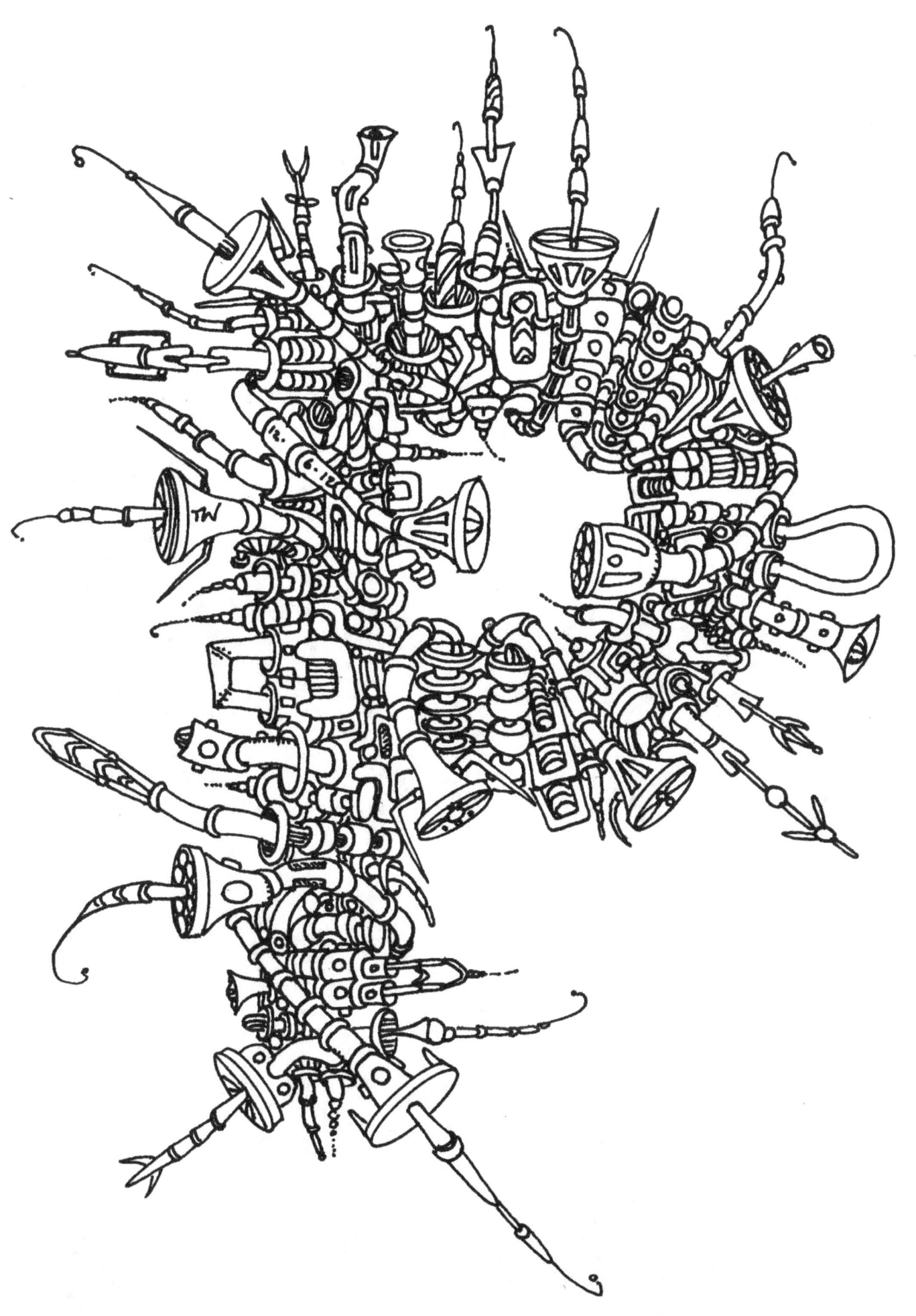

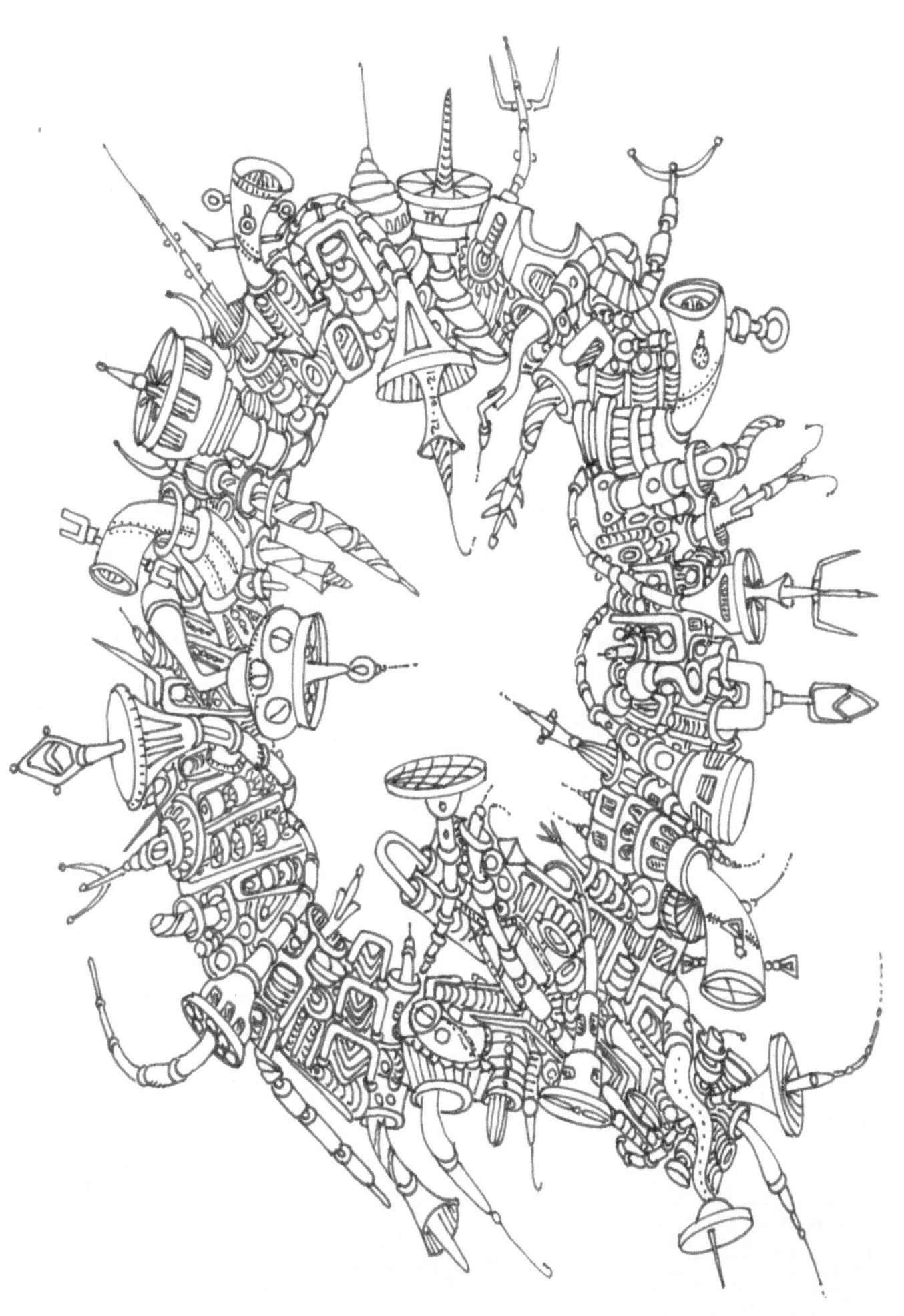

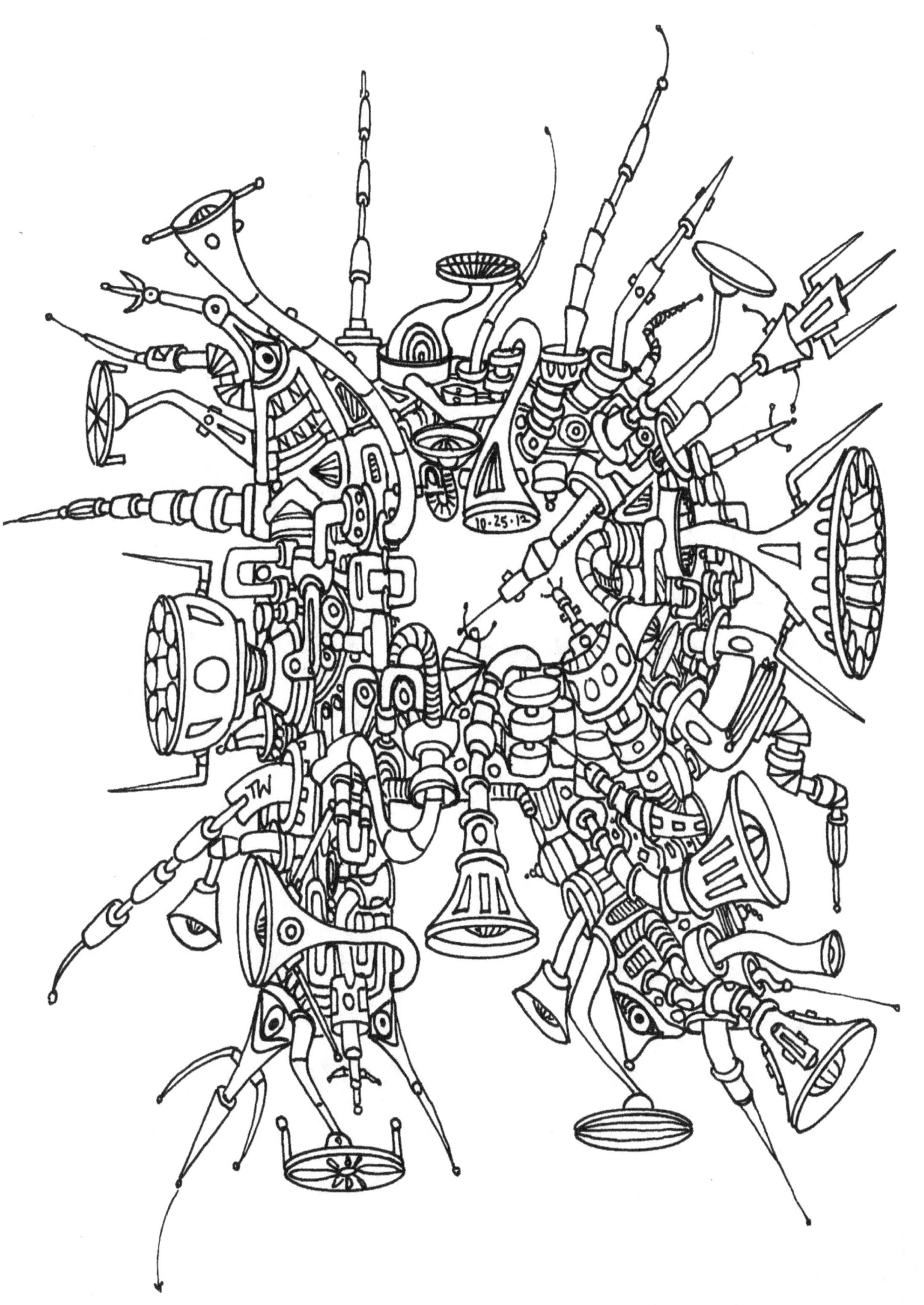

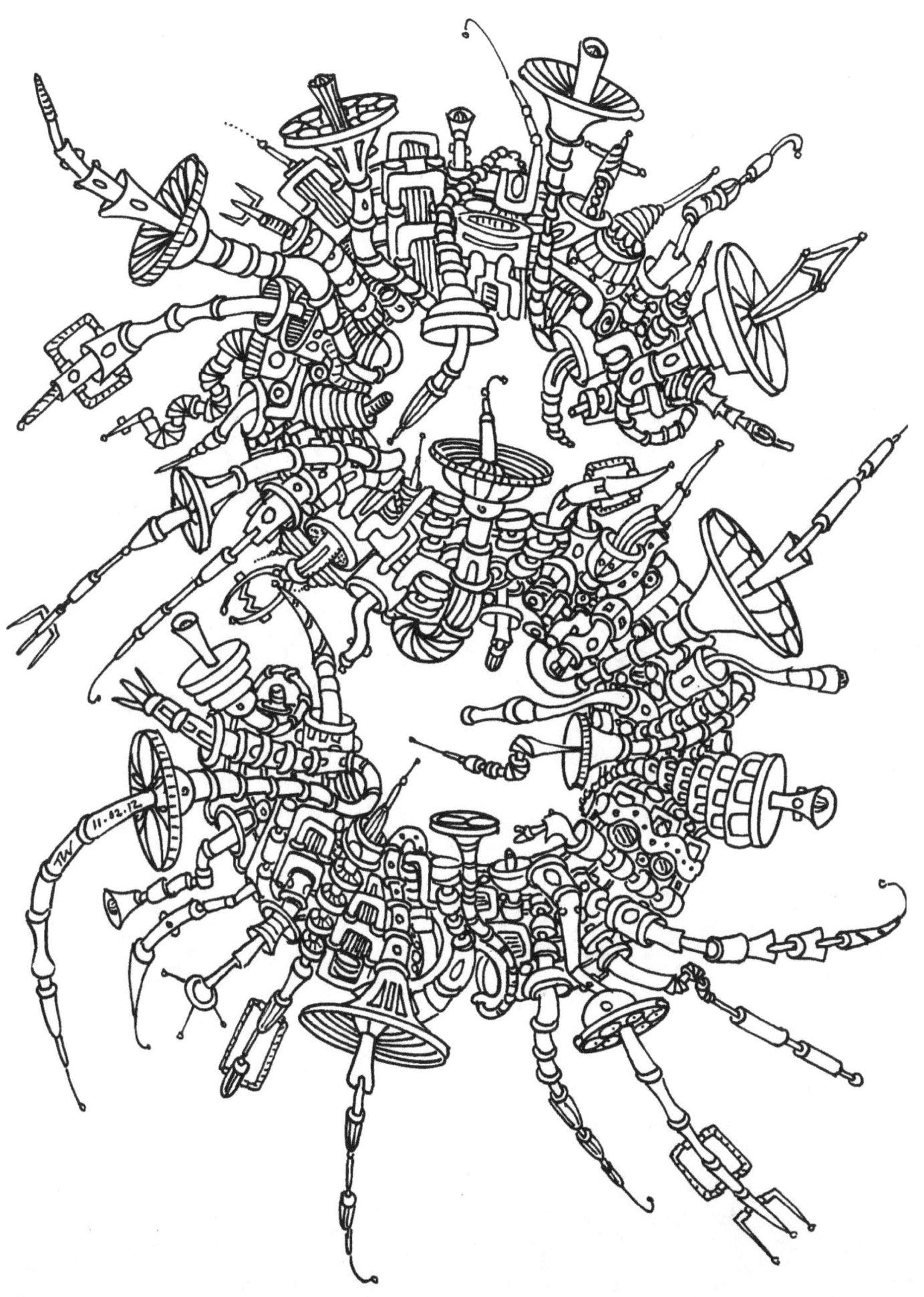

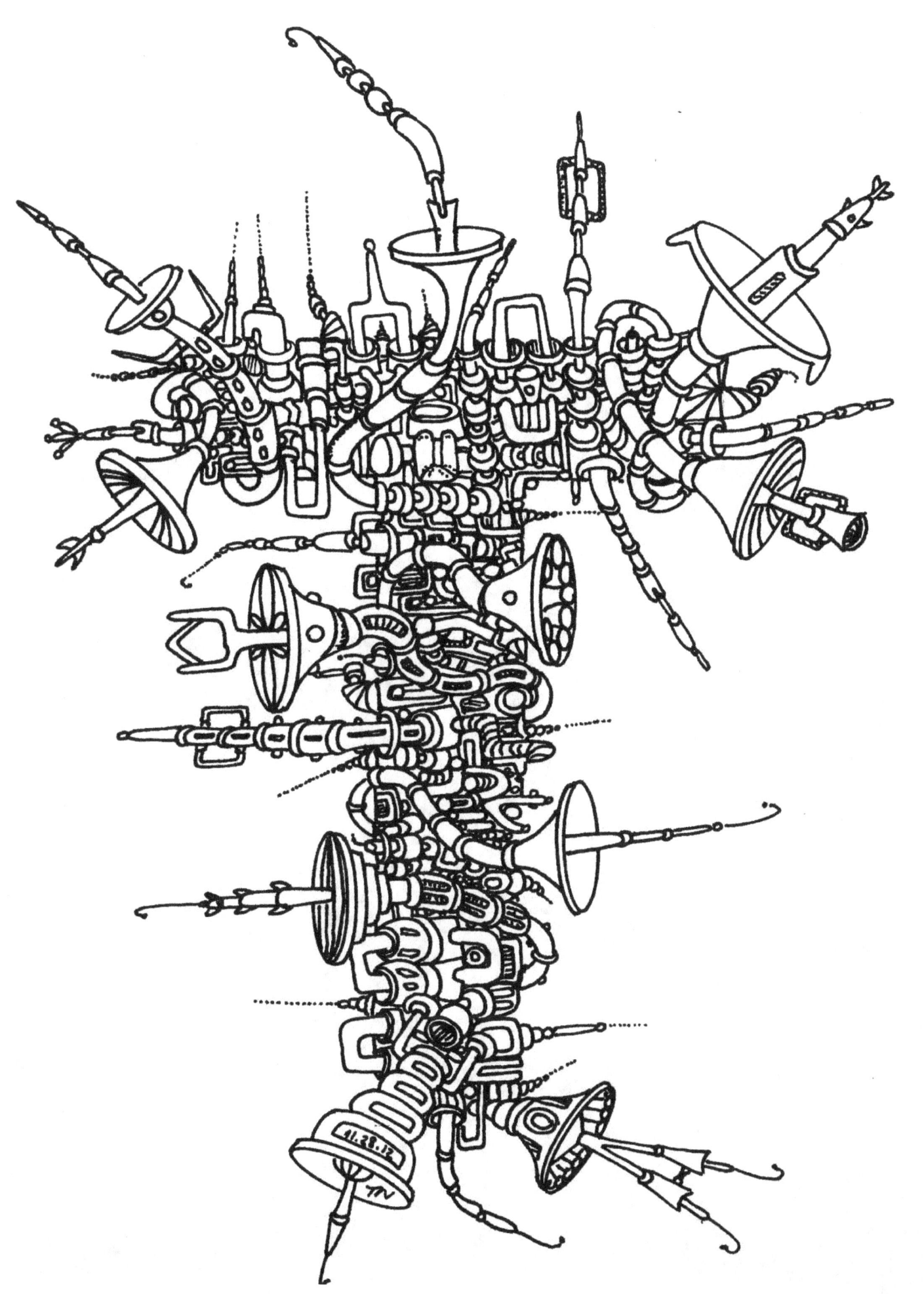

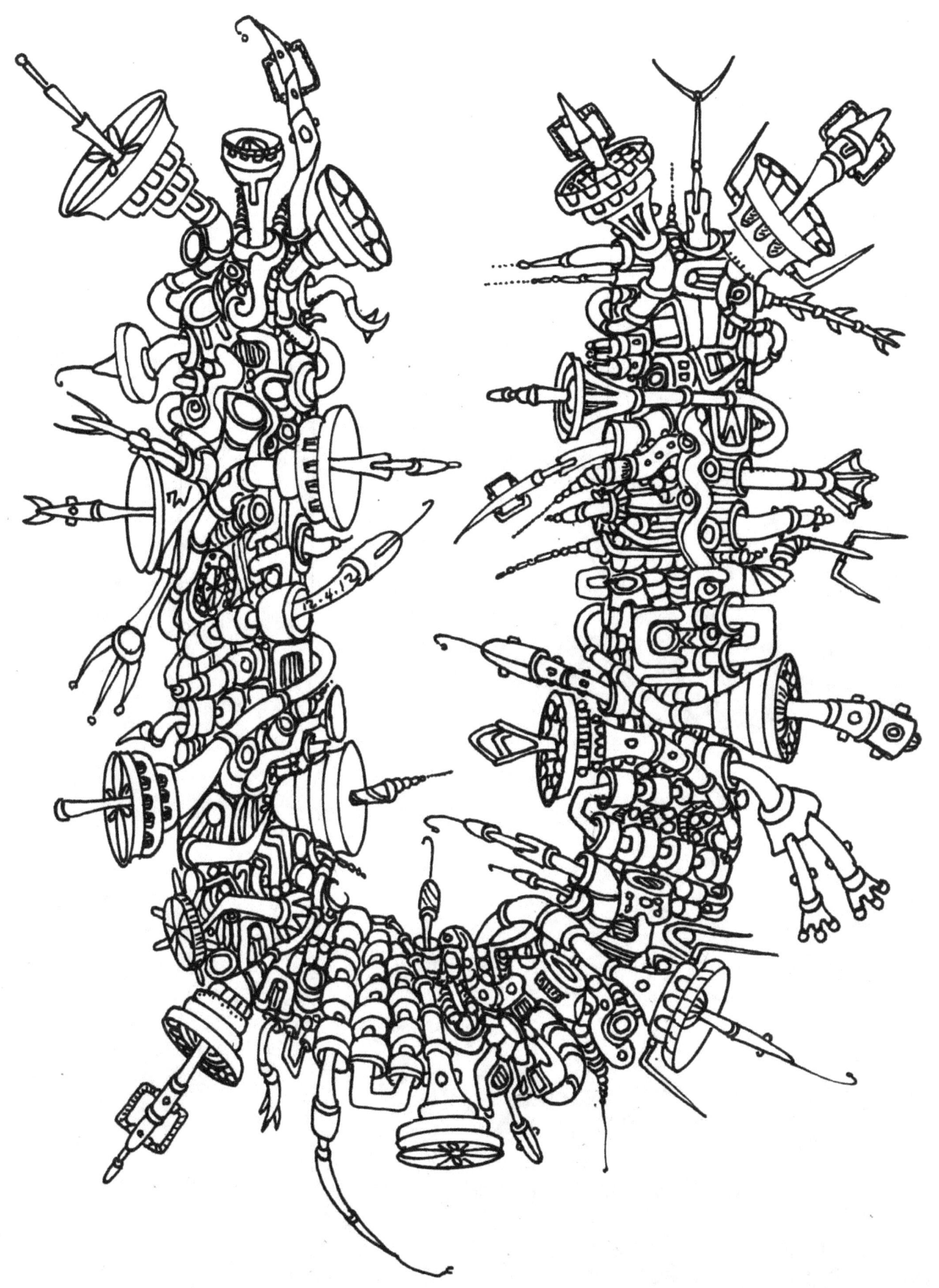

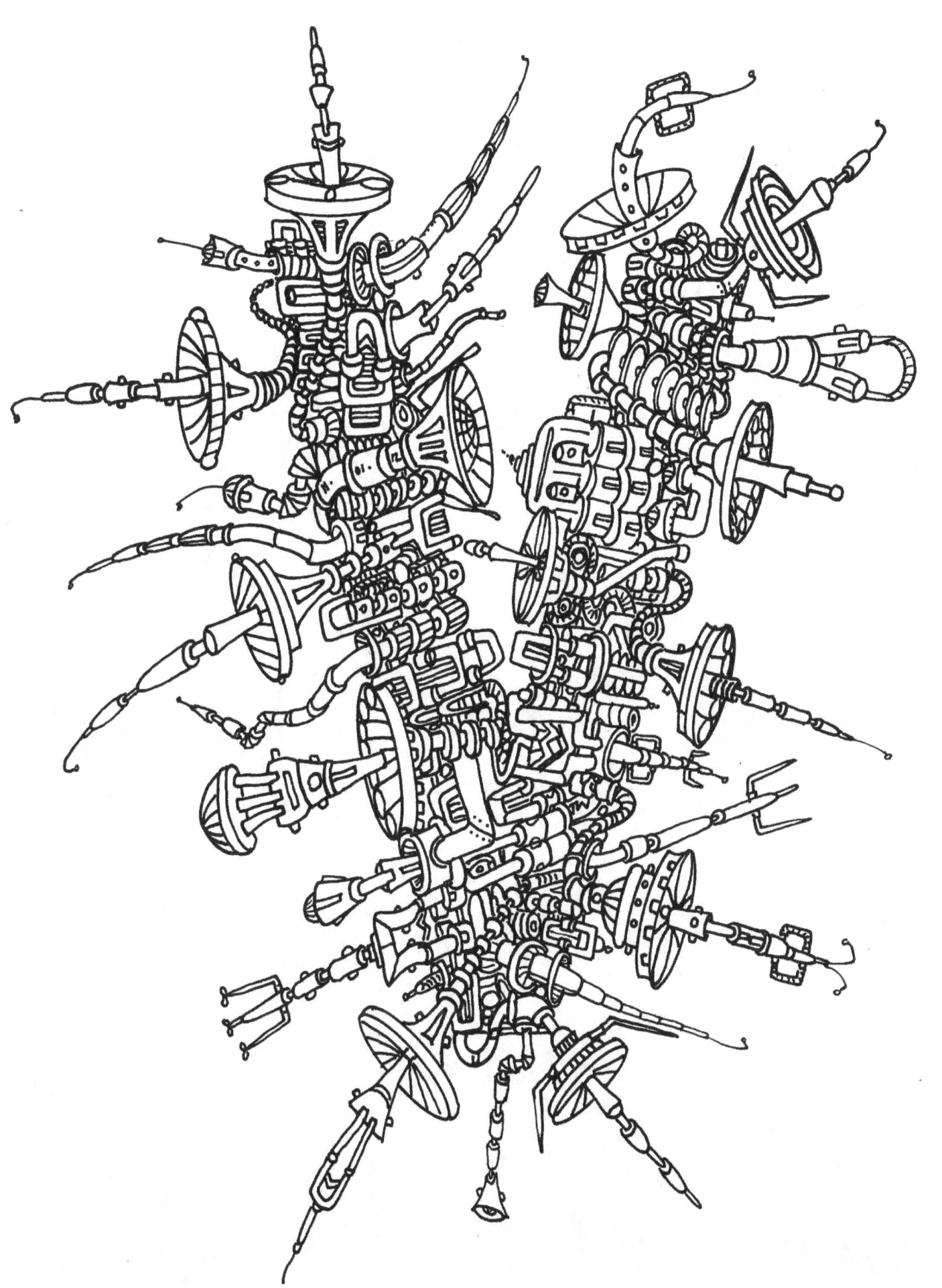

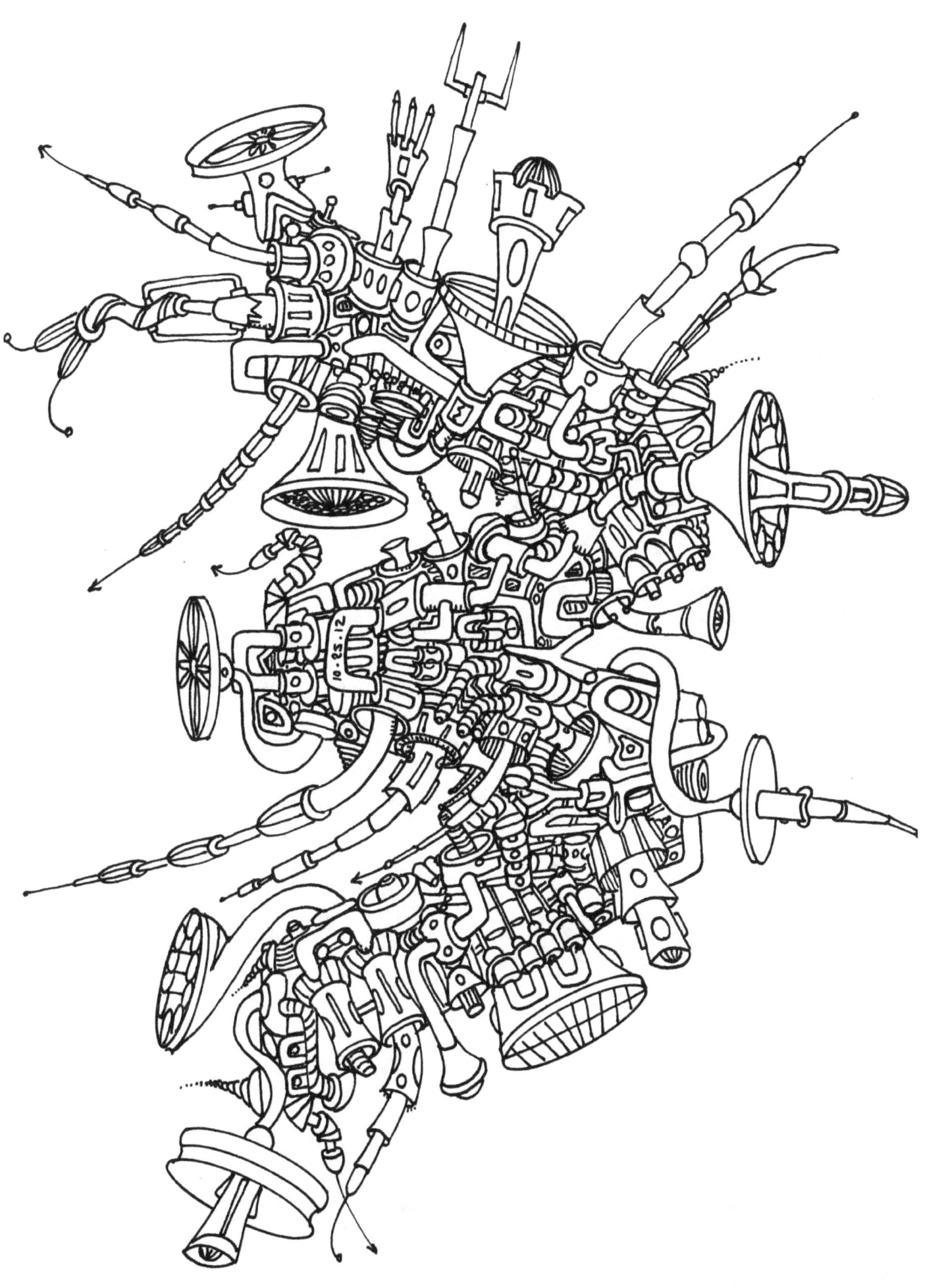

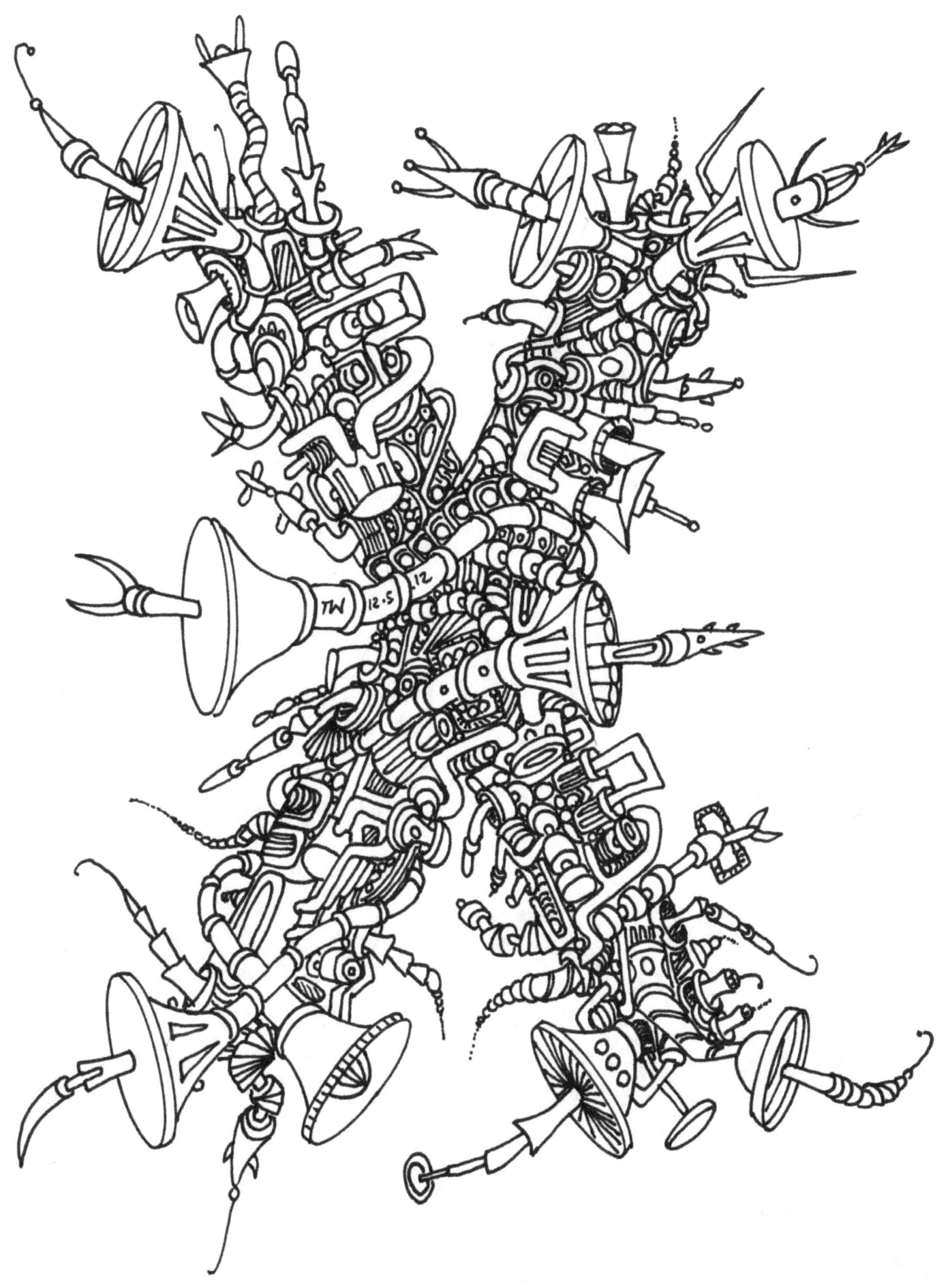

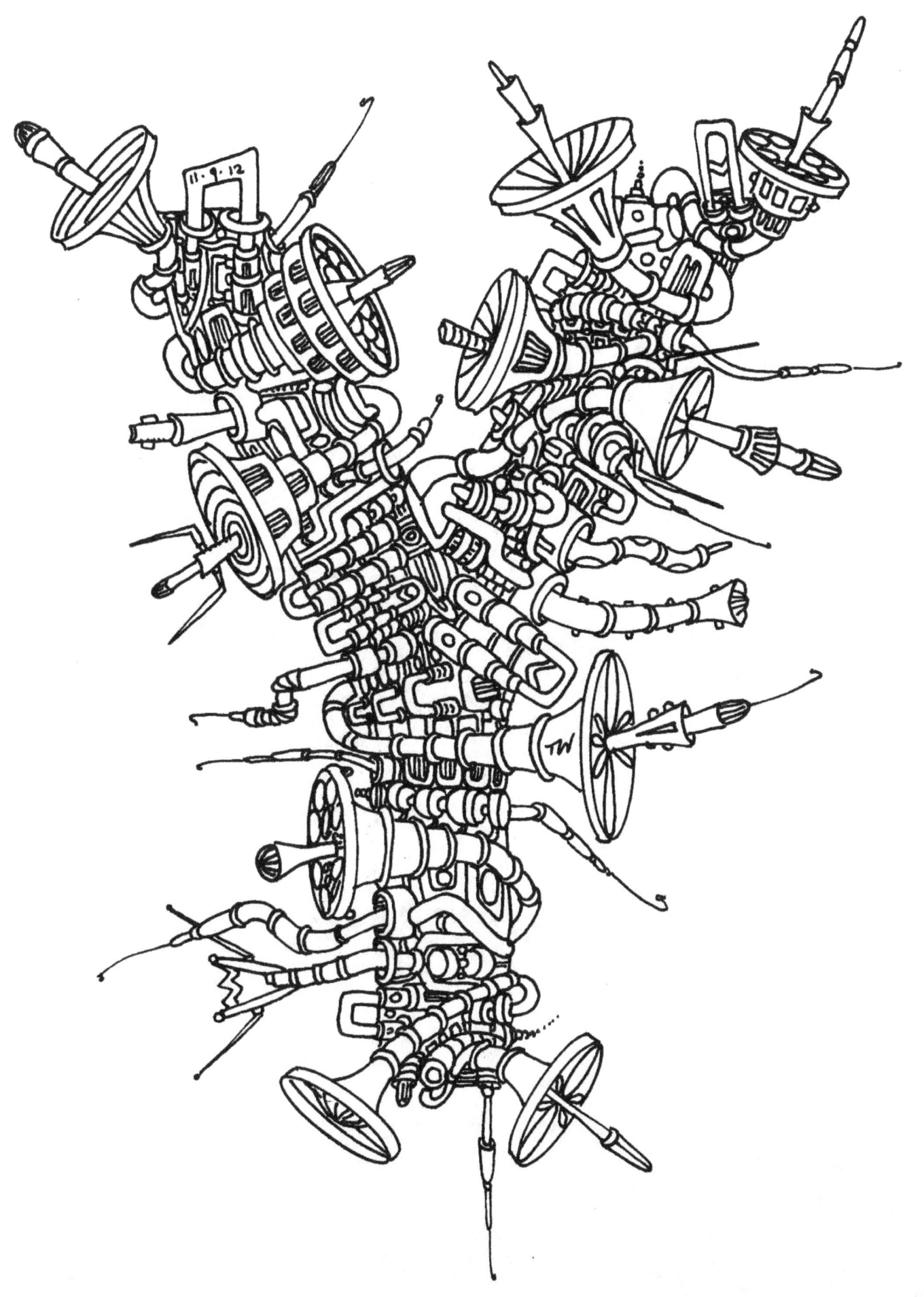

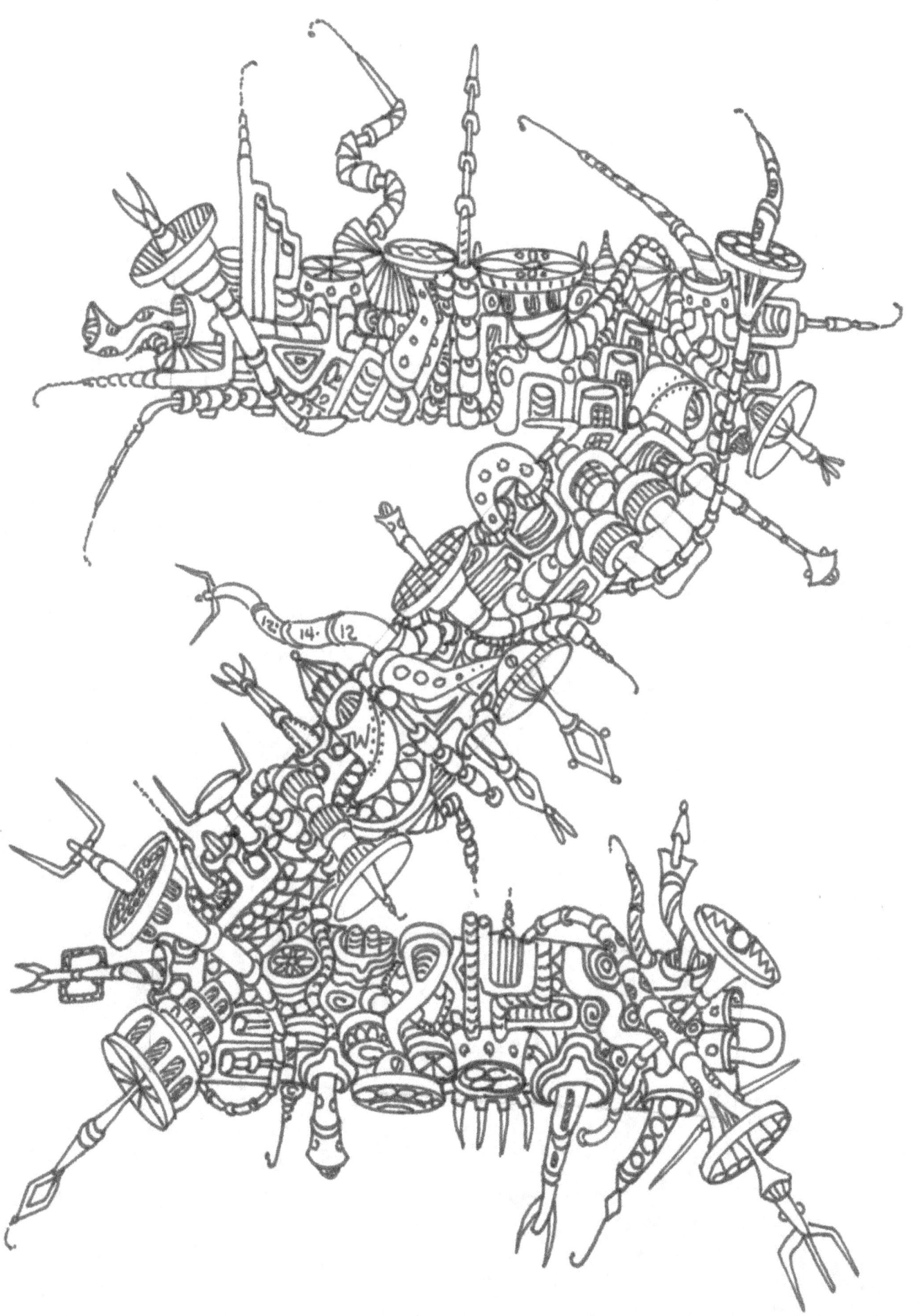

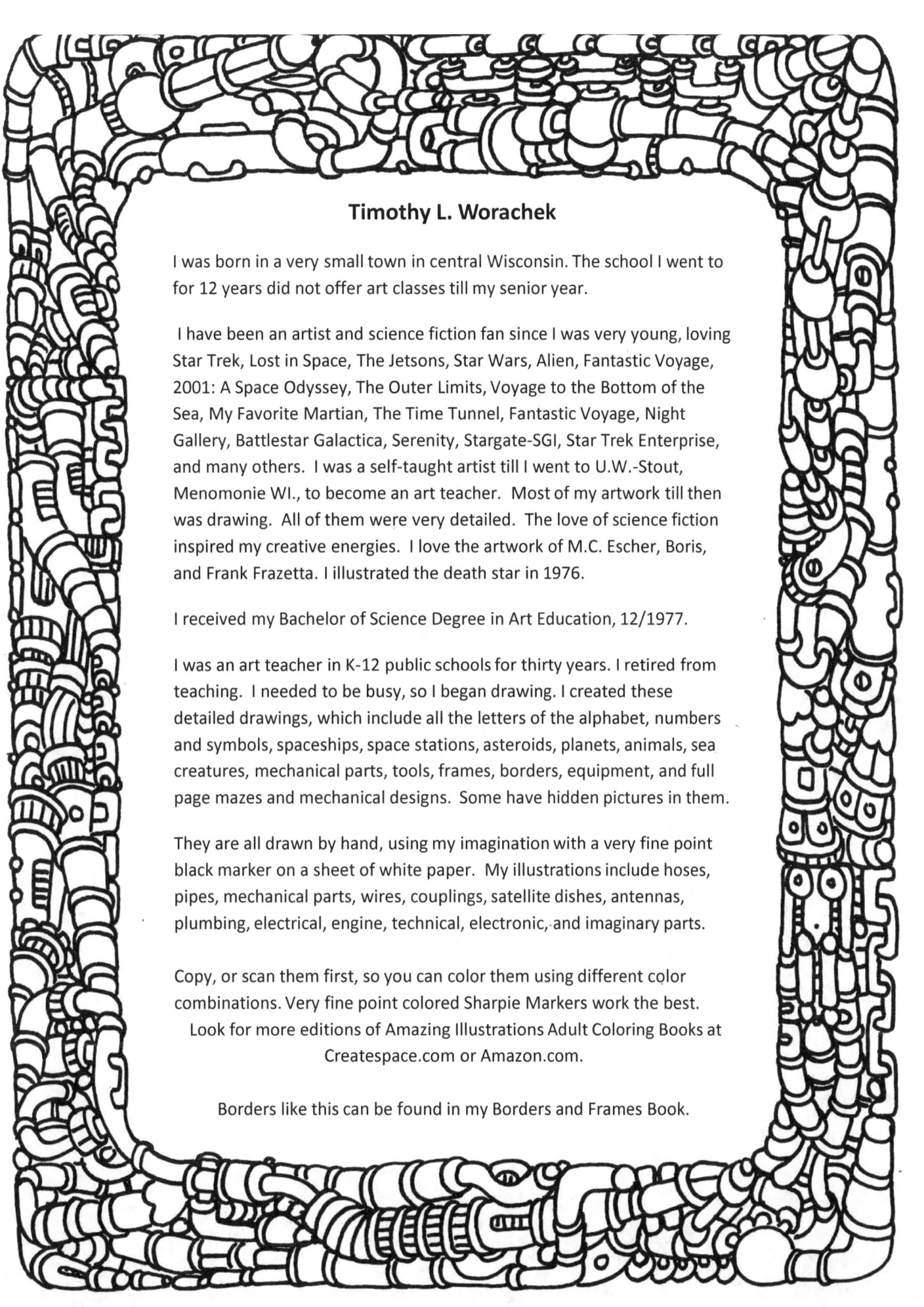

Timothy L. Worachek

I was born in a very small town in central Wisconsin. The school I went to for 12 years did not offer art classes till my senior year.

I have been an artist and science fiction fan since I was very young, loving Star Trek, Lost in Space, The Jetsons, Star Wars, Alien, Fantastic Voyage, 2001: A Space Odyssey, The Outer Limits, Voyage to the Bottom of the Sea, My Favorite Martian, The Time Tunnel, Fantastic Voyage, Night Gallery, Battlestar Galactica, Serenity, Stargate-SGI, Star Trek Enterprise, and many others. I was a self-taught artist till I went to U.W.-Stout, Menomonie WI., to become an art teacher. Most of my artwork till then was drawing. All of them were very detailed. The love of science fiction inspired my creative energies. I love the artwork of M.C. Escher, Boris, and Frank Frazetta. I illustrated the death star in 1976.

I received my Bachelor of Science Degree in Art Education, 12/1977.

I was an art teacher in K-12 public schools for thirty years. I retired from teaching. I needed to be busy, so I began drawing. I created these detailed drawings, which include all the letters of the alphabet, numbers and symbols, spaceships, space stations, asteroids, planets, animals, sea creatures, mechanical parts, tools, frames, borders, equipment, and full page mazes and mechanical designs. Some have hidden pictures in them.

They are all drawn by hand, using my imagination with a very fine point black marker on a sheet of white paper. My illustrations include hoses, pipes, mechanical parts, wires, couplings, satellite dishes, antennas, plumbing, electrical, engine, technical, electronic, and imaginary parts.

Copy, or scan them first, so you can color them using different color combinations. Very fine point colored Sharpie Markers work the best. Look for more editions of Amazing Illustrations Adult Coloring Books at Createspace.com or Amazon.com.

Borders like this can be found in my Borders and Frames Book.

www.ingramcontent.com/pod-product-compliance
Lightning Source LLC
Chambersburg PA
CBHW081301180526
45170CB00007B/2525